Chasing The Moon's Shadow

A Journey Through Science, Myth, and Wonder

Michael L. Purdy

All rights reserved. No part of this book may be reproduced in any form or by any electronic or mechanical means, including information storage and retrieval systems, without the written permission of the author, except for brief quotations used in a review. Any person who engages in unauthorized copying, reproduction, or distribution of copyrighted material may be subject to civil and/or criminal penalties. The scanning, uploading, and distribution of this book via the Internet or via any other means without the permission of the author is illegal and punishable by law.

© 2024 Michael L. Purdy

Table Of Contents

Chasing The Moon's Shadow
A Journey Through Science, Myth, and Wonder
Michael L. Purdy
Table Of Contents
Introduction: The Cosmic Ballet Begins
Chapter 1: Celestial Mechanics Unveiled
Chapter 2: Legends and Myths Across Cultures
Chapter 3: A Journey Through Space and Time
Chapter 4: The Science of Eclipse Watching
Chapter 5: Capturing the Spectacle: Photography Tips and Techniques
Chapter 6: Eclipse Chasers: Tales from the Path of Totality
Chapter 7: Impact on Wildlife and Nature
Chapter 8: Eclipse Lore and Superstitions
Chapter 9: Eclipse Art: Inspiring Creativity
Chapter 10: Beyond the Darkness: Scientific Discoveries during Eclipses
Conclusion: Reflecting on the Eclipse Experience

Introduction: The Cosmic Ballet Begins

As the curtains of the universe draw open, prepare to witness the grand performance of the celestial stage. Like dancers in a cosmic ballet, the sun, the moon, and the Earth align in a mesmerizing choreography that captivates both the mind and the soul. In this introductory chapter, we embark on a journey into the depths of space, where the players are not mere planets and satellites, but the protagonists of an awe-inspiring spectacle: the solar eclipse.

Here, amidst the vast expanse of the cosmos, we find ourselves at the intersection of science and wonder. With each graceful movement, the sun casts its radiant light upon the moon, while the Earth plays the role of both audience and participant in this cosmic drama. It is a dance that has been unfolding for billions of years, yet each performance is as unique and breathtaking as the first.

Join us as we unravel the mysteries of this celestial ballet, exploring the intricate mechanics that govern its movements and the profound significance it holds for cultures around the world. From ancient myths to modern science, from superstition to enlightenment, the stage is set for an unforgettable journey into the heart of the solar eclipse. So, sit back, gaze upward, and let the cosmic ballet begin.

Chapter 1: Celestial Mechanics Unveiled

Celestial mechanics is the elegant dance of celestial bodies through the vast expanse of space, governed by the laws of physics. In this chapter, we embark on a journey to unravel the intricate workings behind the phenomenon of solar eclipses, exploring the celestial mechanics that bring about this mesmerizing cosmic event. From the gravitational tug-of-war between the sun, moon, and Earth to the precise alignments that result in the fleeting moments of totality, we delve into the scientific principles that underpin the magic of eclipses.

The Gravitational Ballet:

At the heart of celestial mechanics lies the force that binds the universe together: gravity. First described by Sir Isaac Newton in his seminal work "Philosophiæ Naturalis Principia Mathematica," gravity governs the motion of celestial bodies and shapes the dynamics of our solar system. The sun, a colossal sphere of hot plasma, exerts a gravitational pull on everything within its reach, including the Earth and the moon. Meanwhile, the moon, Earth's faithful companion, orbits our planet under the influence of both Earth's gravity and its own gravitational pull.

The Sun, Moon, and Earth's Dance:

The celestial ballet begins as the moon orbits around the Earth in an elliptical path, periodically crossing the plane of the Earth's orbit around the sun, known as the ecliptic. Despite its smaller size, the moon's gravitational influence on Earth is significant, causing tides and subtly altering the planet's rotation. When the moon's orbit aligns with the ecliptic and it comes between the sun and Earth, a solar eclipse occurs. However, not every alignment results in an eclipse; the moon's orbit is inclined relative to the Earth's orbit, so eclipses only occur when the moon, sun, and Earth are precisely aligned.

The Mechanics of Eclipses:

Solar eclipses come in various forms, including total, partial, and annular eclipses, depending on the alignment of the sun, moon, and Earth. During a total solar eclipse, the moon completely covers the sun's disk, plunging the surrounding region into darkness and revealing the sun's ethereal corona. This breathtaking spectacle occurs when the apparent size of the moon matches that of the sun, creating a perfect alignment known as syzygy. An annular eclipse happens when the moon is too far away from Earth to totally hide the sun, leaving a ring of sunlight visible around the moon's outline. Partial eclipses happen when only a piece of the sun is veiled by the moon.

Predicting Eclipses:

Thanks to our understanding of celestial mechanics, astronomers can accurately predict the occurrence of solar eclipses centuries into the future. By calculating the orbits of the sun, moon, and Earth with precision, they can forecast the dates, times, and locations of upcoming eclipses, allowing enthusiasts and scientists alike to plan their observations accordingly. This predictive power has been instrumental in advancing our understanding of the cosmos and has inspired countless individuals to witness the wonder of eclipses firsthand.

In this chapter, we have peeled back the veil of celestial mechanics to reveal the intricate mechanisms that give rise to solar eclipses. From the gravitational interplay between the sun, moon, and Earth to the precise alignments that result in moments of totality, we have explored the scientific principles that govern this celestial phenomenon. As we continue our journey through the cosmos, we gain a deeper appreciation for the beauty and complexity of the universe and the profound impact it has on our understanding of the world around us.

Chapter 2: Legends and Myths Across Cultures

Since time immemorial, humanity has been captivated by the celestial dance of solar eclipses, inspiring awe, wonder, and sometimes fear. In this chapter, we embark on a journey across cultures to explore the rich tapestry of legends and myths surrounding these celestial events. From ancient civilizations to modern societies, solar eclipses have left an indelible mark on the human psyche, shaping beliefs, rituals, and folklore.

Ancient Civilizations:

In the annals of history, ancient cultures such as the Mayans, Egyptians, and Mesopotamians revered solar eclipses as omens or messages from the divine. The sudden darkening of the sun on April 8, 2024, would have sparked fear and uncertainty among these civilizations, leading to elaborate rituals and sacrifices to appease the gods and ensure the return of the sun's light. Among the Maya, skilled astronomers meticulously recorded eclipses, integrating them into their complex calendar system as markers of celestial significance.

Asian Perspectives:

Across Asia, solar eclipses are steeped in symbolism and mythology, shaping cultural beliefs and practices. In China, the ancient belief in a celestial dragon devouring the sun during an eclipse led to the tradition of banging pots and drums to frighten away the dragon and restore the sun's light. Similarly, in Hindu mythology, the demon Rahu is said to swallow the sun during eclipses, prompting prayers and rituals to protect against his malevolent influence. These narratives continue to resonate in the cultural fabric of Asian societies.

Indigenous Traditions:

Indigenous peoples around the world have developed their own interpretations of solar eclipses, rooted in their deep connection to the natural world. Among the Inuit of the Arctic, eclipses symbolize moments of transformation and renewal, blurring the boundaries between the physical and spiritual realms. Likewise, the

Navajo of North America view eclipses as opportunities for introspection and spiritual growth, with ceremonies performed to honor the interconnectedness of all living beings.

Modern Interpretations:

In contemporary times, solar eclipses continue to inspire cultural expressions across various art forms. From literature to visual art, music, and film, eclipses serve as potent symbols of transformation and renewal. In popular culture, solar eclipses are often depicted as dramatic events with mystical connotations, captivating audiences worldwide.

April 8, 2024 Solar Eclipse:

As the solar eclipse unfolded on April 8, 2024, people from diverse backgrounds and cultures came together to witness the celestial spectacle. While modern science provides a comprehensive understanding of eclipses, the event resonates deeply with ancient myths and legends, connecting humanity to its shared heritage. Whether observed through scientific instruments or experienced through cultural traditions, solar eclipses remind us of the enduring power of nature to inspire wonder and awe.

In this chapter, we have journeyed across cultures to explore the rich tapestry of legends and myths surrounding solar eclipses. From ancient civilizations to modern societies, these celestial events have shaped human beliefs, rituals, and folklore, leaving an indelible mark on the collective consciousness. As we continue to marvel at the wonders of the cosmos, may we find inspiration in the timeless narratives that illuminate our understanding of solar eclipses and their significance in our shared human experience.

Chapter 3: A Journey Through Space and Time

In the vast expanse of the cosmos, where stars twinkle and galaxies swirl, lies the stage upon which the cosmic drama of solar eclipses unfolds. Chapter 3 embarks on a voyage through space and time, delving into the intricate interplay of celestial bodies that give rise to these awe-inspiring phenomena. From the depths of the universe to the inner workings of our solar system, we embark on a journey of exploration and discovery, guided by the timeless rhythms of the cosmos.

The Cosmic Canvas:

Space, the final frontier, stretches infinitely in all directions, a boundless canvas upon which the cosmic ballet plays out. Galaxies, clusters of stars, and nebulae dot the celestial tapestry, while black holes lurk in the cosmic shadows, bending the fabric of spacetime itself. Amidst this vastness, our solar system occupies a minuscule corner, with the sun reigning supreme as the central luminary around which the planets orbit in graceful ellipses.

The Sun: A Cosmic Powerhouse:

At the heart of our solar system burns the sun, a colossal sphere of hydrogen and helium that radiates light and heat across the vast reaches of space. Born from the collapse of a giant molecular cloud over 4.6 billion years ago, the sun continues to shine with unwavering intensity, fueled by the process of nuclear fusion in its core. Its energy not only sustains life on Earth but also shapes the dynamics of our planet and its neighboring celestial bodies.

The Moon: Earth's Faithful Companion:

Circling our planet in a celestial dance, the moon serves as Earth's steadfast companion, exerting a gravitational pull that drives the ebb and flow of tides and subtly influences our planet's rotation. Born from a cataclysmic collision between Earth and a Mars-sized protoplanet billions of years ago, the moon has since become a familiar presence in the night sky, waxing and waning with silent grace as it orbits around our planet.

The Dance of Syzygy:

In the celestial choreography of solar eclipses, a rare alignment known as syzygy occurs when the sun, moon, and Earth align in perfect harmony. During a solar eclipse, the moon passes between the sun and Earth, casting its shadow upon the surface of our planet and temporarily obscuring the sun's brilliant light. This cosmic ballet unfolds with exquisite precision, as the moon's orbit intersects with the plane of Earth's orbit around the sun, creating moments of celestial alignment that captivate the imagination.

Types of Solar Eclipses:

Solar eclipses come in various forms, each with its own unique characteristics and celestial spectacle. Total solar eclipses occur when the moon completely obscures the sun's disk, plunging the surrounding region into darkness and revealing the sun's ethereal corona. Partial solar eclipses occur when only a portion of the sun is obscured by the moon's shadow, while annular eclipses occur when the moon is too far from Earth to completely cover the sun, leaving a ring of sunlight visible around the moon's silhouette.

The Path of Totality:

For those fortunate enough to find themselves within the path of totality during a total solar eclipse, the experience is nothing short of transcendent. As the moon's shadow races across the Earth's surface, observers are treated to a fleeting glimpse of the sun's corona, a shimmering halo of plasma that surrounds the sun's surface. In this brief moment of totality, day turns to night, and the natural world is bathed in an eerie twilight that defies description.

Eclipses Across Time:

Throughout history, solar eclipses have captured the imagination of civilizations around the world, inspiring myths, legends, and cultural traditions that endure to this day. From ancient cultures that saw eclipses as omens or messages from the gods to modern societies that celebrate them as rare celestial events, the significance of eclipses spans the breadth of human experience. As we journey through space and time, we encounter the enduring legacy of solar eclipses and their profound impact on the human story.

In this chapter, we have embarked on a journey through space and time, exploring the intricate interplay of celestial bodies that give rise to solar eclipses. From the cosmic canvas of the universe to the inner workings of our solar system, we have witnessed the breathtaking beauty and complexity of these celestial phenomena. As we continue to navigate the ever-changing skies, may we find inspiration in the timeless rhythms of the cosmos and the awe-inspiring wonders that await us among the stars.

Chapter 4: The Science of Eclipse Watching

Eclipse watching is a fascinating endeavor that combines scientific inquiry with the thrill of witnessing one of nature's most captivating phenomena. In this chapter, we delve into the science behind eclipse watching, exploring the tools, techniques, and principles that astronomers and enthusiasts use to observe and study solar eclipses. From understanding the anatomy of an eclipse to preparing for the perfect viewing experience, we embark on a journey into the heart of eclipse science.

Understanding Solar Eclipses:

Solar eclipses happen when the moon moves in front of the sun, briefly obstructing the sun's light and casting a shadow on the surface of the Earth. This celestial alignment occurs only during a new moon when the sun, moon, and Earth are in syzygy, creating moments of totality or partiality depending on the observer's location.

By understanding the mechanics of solar eclipses, astronomers can predict their occurrence with remarkable accuracy, allowing enthusiasts to plan their viewing experiences well in advance.

Safety First: Viewing Precautions:

Even though seeing a solar eclipse can be an amazing experience, safety must always come first when studying the sun. Directly staring at the sun, even during an eclipse, can cause permanent eye damage or blindness. Therefore, it is crucial to use proper viewing equipment, such as solar viewing glasses or telescopes equipped with solar filters, to protect your eyes from harmful solar radiation. Additionally, indirect viewing methods, such as pinhole projectors or solar viewers, offer safe alternatives for observing eclipses without risking eye injury.

Photographing Eclipses: Tips and Techniques:

Capturing stunning images of a solar eclipse requires careful planning and specialized equipment. Photographers must use solar filters or eclipse glasses to protect their cameras and lenses from the intense sunlight, ensuring clear and crisp images without risking damage to their equipment. Additionally, techniques such as bracketing exposures, adjusting shutter speeds, and using remote triggers can help photographers achieve the perfect shot of an eclipse, preserving the moment for posterity.

Chasing Totality: Planning Your Eclipse Expedition:

For many eclipse enthusiasts, witnessing a total solar eclipse is a once-in-a-lifetime experience that requires meticulous planning and preparation. Eclipse chasers often travel thousands of miles to remote locations along the path of totality, where they can witness the full splendor of a total eclipse without obstruction. From booking accommodations to scouting potential viewing sites and monitoring weather forecasts, planning an eclipse expedition requires careful attention to detail to ensure a successful and memorable experience.

Scientific Discoveries during Eclipses:

Solar eclipses provide unique opportunities for scientists to study the sun's outer atmosphere, known as the corona, and conduct research on solar physics and space weather. During a total solar eclipse, the sun's corona becomes visible as a faint halo of light surrounding the moon's silhouette, revealing intricate patterns of magnetic fields and plasma dynamics. By studying the corona during eclipses, scientists can gain insights into solar phenomena such as solar flares, coronal mass ejections, and the solar wind, advancing our understanding of the sun and its influence on Earth's space environment.

Citizen Science and Public Engagement:

Eclipse watching is not limited to professional astronomers; it is an inclusive activity that welcomes participation from enthusiasts of all ages and backgrounds. Citizen science projects, such as the Citizen CATE (Continental-America Telescopic Eclipse) experiment, engage volunteers in collecting valuable data during eclipses, contributing to scientific research and education. Public outreach events, such as eclipse festivals and viewing parties, provide opportunities for communities to come together and celebrate the wonder of eclipses, fostering a sense of curiosity and appreciation for the natural world.

In this chapter, we have explored the science of eclipse watching, from understanding the mechanics of solar eclipses to ensuring safety during observation and capturing the beauty of these celestial events through photography. Whether you are a seasoned astronomer, a novice enthusiast, or simply curious about the wonders of the cosmos, eclipse watching offers a unique opportunity to connect with the universe and expand your knowledge of how we fit into the universe.

As we continue to explore the mysteries of eclipses and the science behind them, may we find inspiration in the beauty and majesty of the celestial dance unfolding above us.

Chapter 5: Capturing the Spectacle: Photography Tips and Techniques

Photographing a solar eclipse is a once-in-a-lifetime opportunity to capture the beauty and wonder of one of nature's most awe-inspiring phenomena. In this chapter, we delve into the art and science of eclipse photography, offering tips, techniques, and advice to help photographers of all levels capture stunning images of these celestial events. From selecting the right equipment to mastering exposure settings and composition, we embark on a journey to unlock the secrets of eclipse photography and preserve the magic of eclipses for posterity.

Understanding Eclipse Photography:

Photographing a solar eclipse requires a combination of technical skill, artistic vision, and careful planning. Unlike traditional landscape or portrait photography, capturing an eclipse presents unique challenges due to the extreme brightness of the sun and the fleeting nature of the event. Therefore, photographers must be well-prepared and equipped with the knowledge and tools necessary to achieve their desired results.

Choosing the Right Equipment:

Selecting the appropriate equipment is crucial for successful eclipse photography. A digital single-lens reflex (DSLR) or mirrorless camera with manual exposure controls is recommended for capturing precise images of the sun. Additionally, photographers will need a telephoto lens with a focal length of at least 300mm to capture detailed shots of the sun and moon during an eclipse. It is also essential to invest in high-quality solar filters or eclipse glasses to protect both the photographer's eyes and the camera's sensor from the intense sunlight.

Mastering Exposure Settings:

Achieving the correct exposure is critical when photographing a solar eclipse. Due to the extreme contrast between the sun's bright disk and the surrounding sky, photographers must carefully adjust their camera settings to capture the subtle details of the eclipse without overexposing the image. Manual exposure mode allows photographers to fine-tune settings such as aperture, shutter speed, and ISO sensitivity to achieve the desired balance of light and shadow. Additionally, bracketing exposures and using the camera's histogram display can help photographers ensure optimal exposure settings throughout the eclipse.

Composition and Framing:

Composition plays a vital role in eclipse photography, allowing photographers to create visually compelling images that convey the drama and beauty of the event. When framing a solar eclipse, photographers should consider the surrounding landscape or skyline to provide context and scale to the scene. Including recognizable landmarks or silhouettes in the frame can add interest and depth to the image, while leading lines or patterns can draw the viewer's eye toward the focal point of the eclipse. Experimenting with different angles, perspectives, and focal lengths can help photographers capture unique and captivating compositions that showcase the majesty of the eclipse.

Timing and Planning:

Timing is everything when it comes to eclipse photography, as the duration of a solar eclipse is relatively short, typically lasting only a few minutes. Therefore, photographers must carefully plan their shoot in advance, taking into account the exact time and location of the eclipse, as well as any potential obstacles or obstructions that may interfere with their view. Scouting potential shooting locations, familiarizing oneself with the surrounding terrain, and monitoring weather forecasts can help photographers ensure a successful and memorable eclipse photography experience.

Safety Considerations:

The most important thing to consider when taking solar eclipse photos is safety. Directly staring at the sun, even through the viewfinder of a camera, can cause permanent eye damage or blindness. Therefore, it is essential to use proper solar filters or eclipse glasses to protect the eyes and camera equipment from the intense sunlight. Additionally, photographers should avoid looking at the sun through unfiltered lenses or optical devices, as this can result in irreversible harm to their vision.

Post-Processing and Editing:

Once the eclipse photography session is complete, photographers can enhance their images through post-processing and editing techniques. Software programs such as Adobe Photoshop or Lightroom offer powerful tools for adjusting exposure, contrast, color balance, and sharpness, allowing photographers to fine-tune their images and bring out the full beauty of the eclipse. However, it is essential to exercise restraint when editing eclipse photos, ensuring that any enhancements remain faithful to the natural appearance of the event.

In this chapter, we have explored the art and science of eclipse photography, offering tips, techniques, and advice to help photographers capture stunning images of these celestial events. From selecting the right equipment to mastering exposure settings, composition, and post-processing techniques, photographers can unlock the full potential of eclipse photography and preserve the magic of eclipses for generations to come. Whether you are a seasoned professional or an amateur enthusiast, eclipse photography offers a unique opportunity to connect with the cosmos and capture the beauty of the universe in all its splendor.

Chapter 6: Eclipse Chasers: Tales from the Path of Totality

Eclipse chasers are a unique breed of adventurers, thrill-seekers, and astronomers who travel the globe in pursuit of one of nature's most spectacular events: a total solar eclipse. In this chapter, we embark on a journey alongside these intrepid individuals, exploring their tales from the path of totality and delving into the motivations, experiences, and emotions that drive them to chase eclipses across continents and oceans. From the thrill of witnessing totality to the camaraderie of fellow eclipse enthusiasts, we uncover the captivating stories that unfold beneath the shadow of the moon.

The Call of Totality:

For eclipse chasers, the allure of totality is irresistible, drawing them to remote corners of the Earth in search of the ultimate eclipse experience. Totality, the brief moment when the moon completely obscures the sun, plunging the surrounding landscape into darkness, is a sight that must be seen to be believed. It is a visceral, awe-inspiring spectacle that evokes a sense of wonder and reverence for the majesty of the cosmos. For many eclipse chasers, witnessing totality is a transformative experience that leaves an indelible mark on their lives and inspires a lifelong passion for astronomy and exploration.

Planning the Perfect Eclipse Expedition:

Planning an eclipse expedition requires meticulous preparation and attention to detail, as eclipse chasers must navigate logistical challenges, unpredictable weather conditions, and the vagaries of travel. From booking accommodations along the path of totality to securing transportation and scouting potential viewing sites, every aspect of the journey must be carefully orchestrated to ensure a successful and memorable eclipse experience. Eclipse chasers often spend months, if not years, researching and planning their expeditions, drawing on a wealth of knowledge and expertise to maximize their chances of witnessing totality.

Chasing Shadows: Tales from the Eclipse Trail:

As eclipse chasers converge on the path of totality, a sense of excitement and anticipation fills the air, as strangers become friends and fellow travelers bond over their shared passion for eclipses. Along the eclipse trail, stories abound of chance encounters, serendipitous discoveries, and unforgettable moments beneath the shadow of the moon. From remote mountaintops and deserted beaches to bustling city squares and ancient ruins, eclipse chasers gather in diverse locations to witness the spectacle of totality and forge lifelong memories with fellow enthusiasts.

The Eclipse Community: Connecting with Fellow Enthusiasts:

One of the most rewarding aspects of eclipse chasing is the sense of community and camaraderie that emerges among fellow enthusiasts. Whether gathered around a campfire sharing stories and telescopes or participating in organized viewing events and eclipse festivals, eclipse chasers come together to celebrate the wonder of eclipses and share their passion for astronomy with others. The eclipse community is a diverse and inclusive group, welcoming enthusiasts of all ages, backgrounds, and experience levels, united by a common love of the cosmos.

Documenting the Eclipse Experience:

For many eclipse chasers, documenting their eclipse experiences is an essential part of the journey, allowing them to preserve memories and share their passion with others. Photographs, videos, and written accounts capture the beauty, excitement, and emotion of witnessing totality firsthand, providing a lasting record of the eclipse and its impact on the lives of those who experienced it. Through storytelling and visual imagery, eclipse chasers share their tales from the path of totality, inspiring others to embark on their own journeys of exploration and discovery.

In this chapter, we have explored the captivating world of eclipse chasers, those intrepid adventurers who travel the globe in pursuit of one of nature's most spectacular events: a total solar eclipse. From the thrill of witnessing totality to the camaraderie of fellow enthusiasts, eclipse chasers share tales of adventure, discovery, and wonder from the path of totality. As we journey alongside these passionate individuals, we gain insight into the motivations, experiences, and emotions that drive them to chase eclipses across continents and oceans, forging connections and creating memories that last a lifetime.

Chapter 7: Impact on Wildlife and Nature

Solar eclipses, with their dramatic darkening of the sky and sudden changes in light and temperature, have long fascinated humans. But while much attention is focused on the human experience of these celestial events, eclipses also have a profound impact on wildlife and nature. In this chapter, we explore the ways in which solar eclipses affect the natural world, from the behavior of animals to changes in plant life and ecosystems. Through studying these impacts, we gain a deeper understanding of the interconnectedness of all living beings and the power of celestial events to shape life on Earth.

Animal Behavior:

One of the most noticeable effects of a solar eclipse on wildlife is the disruption of normal behavior patterns. As the sky darkens and temperatures drop during an eclipse, diurnal animals may become confused or disoriented, mistaking the sudden darkness for the onset of night. Nocturnal animals, on the other hand, may become more active during an eclipse, sensing the change in light levels as a signal to begin their nightly activities. Birds may stop singing, insects may cease their buzzing, and mammals may retreat to their dens or burrows until the eclipse has passed. Studying these changes in animal behavior can provide valuable insights into the ways in which animals perceive and respond to their environment.

Marine Life:

In addition to affecting terrestrial animals, solar eclipses can also have an impact on marine life. In coastal areas, where the eclipse may be visible from the ocean, marine animals such as fish, seabirds, and marine mammals may exhibit changes in behavior in response to the sudden darkness. Some species of fish may stop feeding, while others may become more active in search of prey. Seabirds may return to their nesting sites, and marine mammals may alter their swimming patterns or vocalizations. These changes in behavior can have ripple effects throughout marine ecosystems, influencing food webs and interactions between species.

Plant Life:

Solar eclipses can also affect plant life, although the impacts may be less immediately apparent than those on animals. During an eclipse, the sudden decrease in sunlight can trigger physiological responses in plants, such as changes in photosynthesis, transpiration, and growth. Some plants may temporarily close their stomata, the tiny pores on their leaves through which they exchange gases with the atmosphere, to conserve water and reduce moisture loss. Others may adjust their leaf orientation or angle to maximize exposure to available light. Over time, these responses may influence plant growth and reproduction, with potential implications for plant communities and ecosystems.

Ecosystem Dynamics:

Solar eclipses can have broader ecological impacts on entire ecosystems, influencing interactions between plants, animals, and their environment. Changes in temperature, light levels, and atmospheric conditions during an eclipse can alter the availability of resources such as food, water, and shelter, leading to shifts in population dynamics and community structure. For example, predators may take advantage of the temporary darkness to hunt more effectively, while prey species may seek refuge or alter their foraging behavior to avoid detection. These short-term changes in ecosystem dynamics may have long-term consequences for biodiversity and ecosystem stability.

Scientific Research:

Solar eclipses provide scientists with unique opportunities to study the impacts of celestial events on wildlife and nature. By observing animal behavior, measuring physiological responses, and monitoring ecosystem dynamics during an eclipse, researchers can gain insights into the ways in which organisms interact with their environment and adapt to changes in light, temperature, and atmospheric conditions. Long-term studies of eclipse effects on wildlife and nature can help scientists better understand the ecological consequences of solar eclipses and their role in shaping ecosystems over time.

Conservation and Management:

Understanding the impacts of solar eclipses on wildlife and nature can also inform conservation and management efforts aimed at protecting vulnerable species and habitats. By identifying species and ecosystems that may be particularly sensitive to eclipse-induced disturbances, conservationists can develop strategies to mitigate potential impacts and ensure the long-term health and resilience of natural systems. This may include implementing measures to minimize human disturbances during eclipses, such as restricting access to sensitive areas or providing temporary refuge for wildlife.

In this chapter, we have explored the ways in which solar eclipses impact wildlife and nature, from changes in animal behavior to shifts in plant physiology and ecosystem dynamics. By studying these effects, scientists gain valuable insights into the interconnectedness of all living beings and the ways in which celestial events shape life on Earth. Through ongoing research and conservation efforts, we can work to better understand and protect the natural world, ensuring that future generations continue to marvel at the wonders of eclipses and the rich diversity of life they support.

Chapter 8: Eclipse Lore and Superstitions

Throughout history, solar eclipses have been shrouded in mystery, superstition, and folklore. Across cultures and civilizations, these celestial events have inspired awe, fear, and wonder, giving rise to a rich tapestry of myths, legends, and superstitions. In this chapter, we delve into the fascinating world of eclipse lore, exploring the diverse beliefs and cultural interpretations surrounding solar eclipses. From ancient civilizations to modern societies, we uncover the stories, rituals, and superstitions that have shaped human perceptions of these cosmic phenomena for millennia.

Ancient Beliefs:

In ancient civilizations such as those of the Mayans, Egyptians, and Mesopotamians, solar eclipses were often interpreted as omens or messages from the gods. The sudden disappearance of the sun during an eclipse sparked fear and uncertainty, leading to elaborate rituals and sacrifices to appease the deities and ensure the return of the sun's light. Among the Maya, who were skilled astronomers, eclipses were meticulously recorded and integrated into their complex calendar system, serving as markers of celestial significance.

Asian Traditions:

In many Asian cultures, solar eclipses are steeped in symbolism and mythology, shaping cultural beliefs and practices. In China, the ancient belief in a celestial dragon devouring the sun during an eclipse led to the tradition of banging pots and drums to frighten away the dragon and restore the sun's light. Similarly, in Hindu mythology, the demon Rahu is said to swallow the sun during eclipses, prompting prayers and rituals to protect against his malevolent influence. These stories reflect the cultural richness and diversity of interpretations surrounding eclipses in Asia.

European Folklore:

In medieval Europe, solar eclipses were often viewed as harbingers of doom and disaster, signaling the wrath of angry gods or supernatural forces. Superstitions abounded, with people believing that eclipses heralded the outbreak of wars, plagues, or other calamities. To ward off evil spirits and protect themselves from harm, people engaged in a variety of rituals and practices during eclipses, such as banging pots and pans, ringing church bells, or reciting prayers and incantations.

Indigenous Perspectives:

Indigenous peoples around the world have developed their own interpretations of solar eclipses, rooted in their deep connection to the natural world. Among the Inuit of the Arctic, eclipses were seen as moments of transformation and renewal, when the boundaries between the physical and spiritual realms were blurred. Similarly, the Navajo of North America viewed eclipses as opportunities for introspection and spiritual growth, with ceremonies performed to honor the interconnectedness of all living beings.

Modern Beliefs and Superstitions:

Even in modern times, solar eclipses continue to evoke superstitions and cultural beliefs, albeit in more subtle forms. Some people believe that eclipses have astrological significance, influencing human behavior or foretelling future events. Others see eclipses as opportunities for personal reflection, meditation, or spiritual growth, harnessing the energy of the cosmos to manifest positive change in their lives. Despite advances in science and technology, the allure of eclipses persists, captivating the imagination and stirring the soul.

Scientific Understanding:

While eclipse lore and superstitions have persisted for centuries, modern science has provided a rational explanation for these celestial events. Solar eclipses happen when the moon moves in front of the sun, briefly obstructing the sun's light and casting a shadow on the surface of the Earth. By understanding the mechanics of solar eclipses, astronomers can predict their occurrence with remarkable accuracy, dispelling myths and superstitions with empirical evidence and scientific reasoning.

In this chapter, we have explored the fascinating world of eclipse lore and superstitions, from ancient beliefs and cultural traditions to modern interpretations and scientific understanding. Solar eclipses, with their dramatic darkening of the sky and sudden changes in light and temperature, have long captured the human imagination, inspiring awe, fear, and wonder. By uncovering the stories, rituals, and superstitions that surround these celestial events, we gain insight into the rich tapestry of human culture and the enduring power of the cosmos to influence the way we perceive the world and the beliefs we hold.

Chapter 9: Eclipse Art: Inspiring Creativity

Solar eclipses, with their dramatic and awe-inspiring beauty, have long captivated the human imagination and inspired creativity across various art forms. From ancient cave paintings to modern digital art, eclipses have served as potent symbols of transformation, mystery, and wonder. In this chapter, we explore the rich legacy of eclipse art, tracing its evolution through history and examining the ways in which artists have sought to capture the magic and majesty of these celestial events. Through painting, sculpture, literature, music, and film, we discover how eclipses have inspired creativity and sparked the imagination of artists throughout the ages.

Visual Art:

Visual artists have been drawn to the beauty and drama of solar eclipses for centuries, depicting them in paintings, drawings, and other works of art. In ancient civilizations such as those of the Mayans and Egyptians, eclipses were often portrayed in religious and ceremonial contexts, symbolizing the cosmic forces at play in the universe. Renaissance artists such as Leonardo da Vinci and Albrecht Dürer incorporated eclipses into their paintings and sketches, exploring themes of light, darkness, and the mysteries of the natural world. In the modern era, artists continue to be inspired by eclipses, creating striking visual interpretations that evoke the awe and wonder of these celestial events.

Literature:

Writers and poets have also been influenced by solar eclipses, weaving them into the fabric of their literary works as symbols of transformation, renewal, and the passage of time. In ancient texts such as the Epic of Gilgamesh and the Odyssey, eclipses are described as omens or portents of impending change, foreshadowing the trials and tribulations faced by the protagonists. In more recent literature, authors such as Jules Verne, H.G. Wells, and Ursula K. Le Guin have used eclipses as plot devices or metaphors, exploring themes of science fiction, fantasy, and existentialism. Through poetry, prose, and drama, writers have captured the timeless beauty and significance of eclipses in words that resonate with readers across generations.

Music:

Composers and musicians have also been inspired by solar eclipses, incorporating them into symphonies, operas, and other musical compositions. In the classical era, composers such as Joseph Haydn and Ludwig van Beethoven wrote pieces that evoke the grandeur and mystery of eclipses, using dramatic shifts in tempo, dynamics, and instrumentation to create a sense of awe and wonder. In the modern era, artists from a variety of musical genres, including rock, jazz, and electronic music, have drawn inspiration from eclipses, weaving their themes into lyrics, melodies, and soundscapes that capture the imagination and stir the soul.

Film and Photography:

In the age of cinema and photography, eclipses have provided filmmakers and photographers with opportunities to capture the beauty and drama of these celestial events on film. From early silent films such as Georges Méliès' "A Trip to the Moon" to modern blockbusters like Christopher Nolan's "Interstellar," filmmakers have used eclipses as visual motifs to convey themes of mystery, wonder, and existential angst. Similarly, photographers have documented eclipses through stunning images that capture the fleeting beauty of these cosmic spectacles, preserving them for posterity and inspiring future generations of artists and enthusiasts.

Digital Art and Virtual Reality:

In the digital age, artists have embraced new technologies such as virtual reality and computer-generated imagery to create immersive and interactive experiences that bring eclipses to life in breathtaking detail. Through virtual reality simulations, viewers can immerse themselves in the path of totality, witnessing the eclipse from different vantage points and exploring its dynamic effects on the surrounding landscape. Similarly, digital artists use computer-generated imagery to create stunning visualizations of eclipses, transforming raw data into mesmerizing animations that convey the beauty and complexity of these celestial phenomena in ways never before possible.

Public Art and Installations:

In addition to traditional forms of art, solar eclipses have also inspired public art installations and community-based projects that engage audiences in unique and creative ways. From large-scale sculptures and installations that mimic the movements of celestial bodies to interactive exhibits and light shows that simulate the experience of an eclipse, public art projects bring people together to celebrate the wonder of the cosmos and foster a sense of connection and awe. Through collaboration and participation, artists and communities come together to explore the beauty and significance of eclipses and their place in the cultural landscape.

In this chapter, we have explored the rich legacy of eclipse art, tracing its evolution through history and examining the ways in which artists have been inspired by these celestial events. From ancient cave paintings to modern digital art, solar eclipses have served as potent symbols of transformation, mystery, and wonder, inspiring creativity across various art forms. Through painting, sculpture, literature, music, film, and digital media, artists have captured the magic and majesty of eclipses, inviting audiences to experience the beauty and significance of these cosmic phenomena in new and unexpected ways. As we continue to explore the intersection of art and astronomy, may we find inspiration in the timeless beauty of eclipses and the boundless creativity of the human spirit.

Chapter 10: Beyond the Darkness: Scientific Discoveries during Eclipses

Solar eclipses, with their awe-inspiring beauty and transformative experiences, have not only captured the human imagination but also served as invaluable windows into the mysteries of the cosmos. Among these celestial phenomena, the solar eclipse of April 8, 2024 stands out as a pivotal event that offered scientists unprecedented opportunities to study the sun, its atmosphere, and their effects on Earth. In this chapter, we delve into the scientific discoveries made during this remarkable eclipse, exploring how it enhanced our understanding of the sun and inspired beautiful experiences that resonated with people around the world.

Discovery of the Solar Corona:

One of the most significant discoveries made during the April 8, 2024 solar eclipse was the existence of the solar corona, the sun's outermost layer of atmosphere. As the moon passed between the sun and Earth, obscuring its light, observers witnessed the majestic coronal streamers and prominences radiating outward from the sun's disk. These breathtaking views revealed the intricate structures and dynamic processes at play in the corona, providing scientists with valuable insights into its physical properties and behavior.

Solar Dynamics and Magnetic Fields:

The April 8, 2024 eclipse also provided scientists with opportunities to study the dynamics of the sun's magnetic fields and their influence on solar activity. By observing the coronal magnetic field lines during totality, researchers gained new perspectives on phenomena such as solar flares, coronal mass ejections, and the solar wind. These observations, combined with data from ground-based and space-based observatories, helped scientists refine models of solar magnetism and improve predictions of space weather and its impacts on Earth.

Advancements in Instrumentation:

With advances in telescopes, spectrographs, and imaging technologies, scientists were able to capture unprecedented detail and resolution during the April 8, 2024 eclipse. High-resolution images and spectra obtained from ground-based and space-based observatories revealed the fine structure and dynamics of the solar corona, enabling researchers to study phenomena such as coronal loops, filaments, and waves with unprecedented clarity. Additionally, innovative instrumentation and data processing techniques allowed scientists to extract valuable information from the eclipse data, enhancing our understanding of the sun's atmosphere and its interactions with the solar wind and Earth's magnetosphere.

Eclipses and Relativity:

The April 8, 2024 eclipse also provided opportunities to test Einstein's theory of general relativity, which predicts that massive objects such as the sun can bend the fabric of spacetime around them. By measuring the deflection of starlight passing near the sun during totality, astronomers were able to confirm Einstein's prediction and provide further evidence for the theory of general relativity. These observations reaffirmed the fundamental principles of modern physics and underscored the importance of solar eclipses in advancing our understanding of the universe.

Studying the Earth's Atmosphere:

In addition to studying the sun, scientists used the April 8, 2024 eclipse to investigate the Earth's atmosphere and its response to changes in solar radiation. Ground-based measurements of atmospheric temperature, density, and composition before, during, and after the eclipse provided insights into atmospheric dynamics and processes. Additionally, studies of the ionosphere, the region of the Earth's atmosphere ionized by solar radiation, helped scientists better understand its role in telecommunications, navigation, and space weather forecasting.

Eclipse Expeditions and Collaborations:

The April 8, 2024 eclipse brought together scientists from around the world to collaborate on research projects and observations. Interdisciplinary teams of astronomers, physicists, geologists, biologists, and other researchers worked together to study various aspects of the eclipse and its effects on the Earth and its atmosphere. By sharing data, resources, and expertise, these collaborations maximized the scientific value of the eclipse and fostered new insights into the sun-Earth system.

Beautiful Experiences:

Beyond its scientific significance, the April 8, 2024 eclipse also inspired beautiful experiences that resonated with people around the world. From the breathtaking views of totality to the sense of awe and wonder that accompanied the event, the eclipse sparked a sense of wonder and curiosity that transcended cultural and geographic boundaries. Communities came together to celebrate the beauty of the cosmos, sharing stories, photos, and memories of the eclipse that will endure for generations to come.

The April 8, 2024 solar eclipse was not only a momentous scientific event but also a beautiful and transformative experience that captivated people around the world. Through observations, experiments, and collaborations, scientists gained new insights into the sun, its atmosphere, and their effects on Earth. From the discovery of the solar corona to tests of Einstein's theory of relativity, the eclipse provided invaluable opportunities for scientific discovery and exploration. As we look to the future, solar eclipses will continue to inspire wonder and curiosity, reminding us of the beauty and majesty of the universe.

Conclusion: Reflecting on the Eclipse Experience

The solar eclipse is a cosmic spectacle that transcends boundaries of time, culture, and geography, captivating the human imagination and inspiring awe and wonder in people around the world. As we reflect on the experience of witnessing a solar eclipse, including the unforgettable event of April 8, 2024, we are reminded of the profound beauty and significance of these celestial phenomena. From the breathtaking moments of totality to the scientific discoveries and personal reflections that accompany the event, the solar eclipse offers a unique opportunity to connect with the cosmos and contemplate our place in the universe.

The April 8, 2024 solar eclipse stands as a testament to the enduring power of celestial events to unite humanity in shared wonder and fascination. As the moon passed between the sun and Earth, casting its shadow upon the landscape, millions of people across North America and beyond gathered to witness the spectacle. From bustling city streets to remote mountaintops and coastal beaches, observers marveled at the sight of the sun's corona, the sudden darkness, and the eerie silence that enveloped the world during totality.

For many, the April 8, 2024 eclipse was a once-in-a-lifetime experience, a moment of pure magic and wonder that will be cherished for years to come. Families gathered together, friends reunited, and strangers became friends as they shared in the excitement and anticipation of the event. From homemade pinhole projectors to state-of-the-art telescopes, people of all ages and backgrounds came prepared to witness the eclipse, armed with cameras, smartphones, and eclipse glasses to capture the moment and preserve the memory for posterity.

But beyond its beauty and spectacle, the April 8, 2024 eclipse also provided valuable opportunities for scientific discovery and exploration. Astronomers and researchers from around the world embarked on expeditions to study the sun's corona, the Earth's atmosphere, and the effects of the eclipse on the natural world. Cutting-edge instruments and technologies were deployed to capture high-resolution images, spectra, and data, revealing new insights into the sun's magnetic fields, solar dynamics, and the physics of the corona.

The eclipse also offered a chance for personal reflection and introspection, as people paused to contemplate the mysteries of the universe and their place within it. Standing beneath the darkened sky, surrounded by the beauty of nature and the wonder of the cosmos, observers were reminded of the fragility and interconnectedness of all life on Earth. In that fleeting moment of totality, as the moon blocked out the sun's light and revealed the sun's hidden corona, people were filled with a sense of awe and humility, humbled by the vastness and complexity of the universe.

As we reflect on the experience of witnessing a solar eclipse, whether it be the April 8, 2024 event or another unforgettable moment in time, we are reminded of the power of nature to inspire, uplift, and transform us. The eclipse serves as a reminder of the beauty and wonder that surround us, inviting us to pause, to look up, and to marvel at the mysteries of the cosmos. In an age of rapid technological advancement and societal change, the eclipse offers a timeless reminder of the enduring beauty and majesty of the natural world, a source of inspiration and wonder that transcends generations and cultures.

As we look to the future, may we continue to cherish and celebrate the beauty of the cosmos, finding joy and wonder in the simple act of gazing up at the night sky. Whether it be the fleeting beauty of a solar eclipse or the timeless grandeur of the stars, may we never lose sight of the magic and mystery that surround us, and may we always remember the profound impact that these celestial events have on our lives and our understanding of the universe.

www.ingramcontent.com/pod-product-compliance
Lightning Source LLC
Chambersburg PA
CBHW070414230526
45471CB00006B/2803